Table of Contents

Executive Summary 1

Chapter 1 2

Introduction 2

    Background 2

    Overview of the Response and Recovery Process 3

    Capability Gaps 4

    Roadmap Development Process 4

    Purpose and Scope 5

Chapter 2 7

Capability Goals to Support Critical Response and Recovery Decisions 7

    Goal 1: Characterize the extent of the incident to reduce exposure and save lives. 7

    Goal 2: Effectively communicate to reduce the impacts of a biological incident. 9

    Goal 3: Accurately assess risk of exposure and risk of infection. 12

    Goal 4: Reduce the risk of exposure and/or infection. 13

    Goal 5: Manage biological waste following a catastrophic incident. 15

Conclusions 18

Appendix I: Definitions 19

Executive Summary

A catastrophic biological incident could threaten the Nation's human, animal, plant, environmental, and economic health, as well as America's national security. Such an event would demand a quick and effective response in order to minimize loss of life and other adverse consequences and, in the case of suspected criminal activity or terrorism, to thwart ongoing activity and prevent follow-on attacks. But response and recovery from a catastrophic biological incident is not a simple, formulaic process. Rather, it is a continuous process of data and information collection, evidence-based review, and decision making, all leading to an informed and constantly evolving series of critical and coordinated actions. Moreover, the response and recovery process involves the integration and coordination of data and capabilities from many different sectors, including public health, law enforcement, waste management, infrastructure management, and transportation. Strategic science and technology (S&T) investments are essential to provide the information that can support evidence-based operational decisions and strengthen response-and-recovery efforts. ***This report categorizes key scientific knowledge gaps, identifies technology solutions to these gaps, and prioritizes research areas that will enable government at all levels to make decisions more effectively during the response to, and recovery from, biological incidents.***

The prioritized, near-term objectives and broader, long-term goals presented in this report constitute a roadmap for use by Federal departments and agencies to coordinate their research and development (R&D) activities. The primary near-term objectives whose timely achievement this roadmap aims to facilitate are:

- Establish the location(s) of the confirmed biological agent in the environment;
- Develop reliable estimates of risk of exposure for a multitude of environments, matrices, and conditions associated with wide-area release scenarios;
- Develop reliable estimates of risk to humans, animals, and plants through various exposure and transmission routes;
- Develop risk reduction strategies, including decontamination, waste management, contaminant control, and reaerosolization control, for a variety of biological threats and scenarios;
- Evaluate population infection prevention measures (e.g., quarantine, isolation, and social distancing) used to reduce incident impact and develop a strong scientific basis for recommending these measures; and
- Use risk communication research to guide development of appropriate messages and dissemination means to stakeholders, including decision makers, first responders, the public, and the media.

Coordination of R&D agendas among Federal departments and agencies will reduce duplication of effort and enhance efficiencies as the Nation enhances its capacity to prevent, protect against, mitigate, respond to, and recover from catastrophic biological incidents.

Chapter 1

Introduction

"Just as we step up our ability to prevent an attack, we must also bolster our capacity to protect against — and respond to — the threats that may come. When it comes to bioterror, this can mean the difference between a contained incident and a catastrophe. That's why we need to invest in new vaccines, to reduce the risk posed by those who would use disease as a weapon. That's why we must develop the technology to detect attacks and to trace them to their origin, so that we can react in a timely fashion. And to care for our citizens who are infected, we must provide our public health system across the country with the surge capacity to confront a crisis. Making these changes will do more than help us tackle bioterror — it will create new jobs, support a healthier population, and improve America's capability to respond to any major disaster."

—President Barack Obama (July 16, 2008)

Background

A biological incident may be caused by a naturally-occurring outbreak with a human, plant, or animal pathogen, such as the 2009 H1N1 pandemic; the deliberate dissemination of pathogens, such as the 2001 U.S. anthrax attacks; or the accidental release of biological agents, such as the 2007 release of foot-and-mouth disease virus from a laboratory in the United Kingdom. Such events can have devastating impacts on public, animal, or plant health, the economy, critical infrastructure, military readiness, and public confidence. Effective response to and recovery from a catastrophic biological incident can mitigate all those risks, but will depend upon rapid, sound decisions being made by response personnel and government officials.

In working toward that goal, the Secretaries of the U.S. Departments of Health and Human Services, Agriculture, Interior, Defense, Commerce, and Homeland Security, as well as the Administrator of the Environmental Protection Agency and the U.S. Attorney General, with the support of other Federal partners, are guided by the National Response Framework[1] and specific Presidential Policy Directive-8[2] frameworks to support local authorities in making critical decisions to enable effective response and recovery from a biological incident.

To ensure the United States is fully able to mount an effective response to and recovery from a catastrophic biological incident, the Subcommittee on Biological Defense Research and Development (BDRD), under the Committee on Homeland and National Security of the National Science and Technology Council, chartered the Biological Response and Recovery Science and Technology (BRRST) Working Group (WG) to assess biological incident response and recovery capabilities and recommend a way forward towards addressing scientific knowledge or technological gaps to improve decision making.

---

[1] Federal Emergency Management Agency, *National Response Framework, Second Edition*, Department of Homeland Security, May, 2013.

[2] *Presidential Policy Directive / PPD-8: National Preparedness* March 30, 2011.

Overview of the Response and Recovery Process

Response to and recovery from a catastrophic biological incident requires a continuous and coordinated process of data and information collection, review, and decision making that results in a series of critical and coordinated actions (Figure 1). Science and technology (S&T) provide the knowledge and tools for effective response and recovery operations. Throughout the response and recovery process, decisions depend on what is known about the agent's transmission dynamics, chain of infection, and other data that may change or evolve as the incident matures and the response unfolds. Response includes those capabilities necessary to save and sustain lives; mitigate human, environmental, plant, and animal health impacts; stabilize the incident; protect property and the environment; meet basic human needs after an incident has occurred; and, in the case of suspected Federal crimes or terrorism, thwart continued activity and prevent follow-on attacks. The ability to make sound decisions in the first minutes, hours, and days following an incident can make a significant difference in lives saved, extent of the spread of disease, and duration of the overall recovery.

Key elements of an effective biological response and recovery are described in the 2009 Draft document, "Planning Guidance for Recovery Following Biological Incidents"[3]. Briefly, the first phase of activities includes detection and confirmation of a biological incident, followed by notification/early warning and first response. Decisions to begin notification procedures and initiate communications regarding an incident require confidence in the identification and confirmation process, as well as guidance on operational coordination and communications strategies for disseminating information about the incident to appropriate authorities and the public.

Notification/early warning procedures provide situational awareness by confirming for proper authorities and the public that a biological incident has occurred. First response is a series of decisions and actions immediately following notification that aim to effectively control, contain, investigate, and mitigate the effects of a biological incident. First response may include initial site containment, environmental sampling and analysis, and public health activities such as treatment of potentially exposed persons and industry engagement. It may also include a law enforcement response, tactical and technical operations, designation of crime scenes, and related activities.

Restoring basic services and supporting the transition to recovery are also critical elements of a timely response to a biological incident. Restoration and recovery encompass the process of returning a community to a state of normality after a disastrous biological incident and requires, among other elements, ongoing characterization of the environment to determine health risks. Efforts to support restoration and, as appropriate, re-occupancy following a biological incident include elements of communicating the risk associated with returning an area to normal use and how best to reduce risk of infection or exposure through remediation processes. Federal, state, or local public health officials, government departments and agencies, and/or property owners (depending on site-specific jurisdictional authorities) make final decisions on clearance.

---

[3] Subcommittee on Decontamination Standards and Technology Committee on Homeland and National Security, *Draft Planning Guidance for Recovery Following Biological Incidents*, National Science and Technology Council, May 2009.

Environmental stability of biological pathogens varies greatly among various organisms, and restoration, clearance, and re-occupancy can range from relatively uncomplicated processes to large-scale decontamination processes.

Capability Gaps

While substantial progress has been made to prevent, detect, respond to, and recover from natural, accidental, and intentional outbreaks, additional focused S&T efforts that leverage investments by Federal departments and agencies and integrate knowledge from advancements in the biological sciences are required to fill critical capability gaps that currently hamper decision makers. Among the numerous factors that can undermine an effective response to an incident are the lag between the incident occurrence and its detection and/or confirmation; lack of understanding of agent spread, transport, and persistence in the environment; lack of clarity about the actual impacted or contaminated area; and uncertainty due to technical limitations and gaps in knowledge and information to drive decision making. Additionally, biological contamination with certain agents presents unique remediation challenges because of the ability of the agent to infect and replicate in a host and/or persist or propagate and thrive in the environment. For example, periodic natural outbreaks of anthrax in animals throughout the world demonstrate the persistence and transport of the organism in the environment. Likewise, seasonal outbreaks of influenza show a persistence of the causative virus in human and animal reservoirs. Biological structure, metabolic characteristics, and natural history of biological agents in conjunction with the physical and climatic conditions in the surrounding environment define the survival rate and hence the fate of the agent outside of its host or hosts. Changing conditions affect the presence and persistence of an agent in the environment, which influences the risk posed and drives the choice of risk-reduction strategies. Therefore, response and recovery processes, including a variety of risk-reduction strategies, are dependent upon a comprehensive understanding of the agent and the human and environmental contexts in which it exists.

Roadmap Development Process

To define the path forward outlined in this *Roadmap*, the BRRST WG developed a working document that described the decisions that first responders or government officials would need to make following a biological incident, what questions the decision maker might ask, and, of those questions, which could be addressed with scientific information or technological capabilities. While the framework established in the draft "Planning Guidance for Recovery Following Biological Incidents" focuses only on remediation/cleanup recovery operations after contamination with *Bacillus anthracis*, the causative agent of anthrax, that framework serves as a starting point for describing the decisions that need to be made in response to biological incidents in general. Based on the phases and activities outlined in that framework, the BRRST WG developed a list of major (high-impact) decisions that would need to be made during each phase of response and recovery (Figure 1). (The decisions and activities presented in Figure 1 are nominally categorized according to their relative timing within the response and recovery timeline but the timing of their implementation can vary—e.g., some limited decontamination operations and waste generation may begin with the first response activities and implementing public messaging generally occurs throughout the entire process).

The resultant working document was used to guide discussions and to collect Federal department and agency information on current and planned programmatic activity with the potential to address the questions answerable by S&T. Review of submitted information on Federal activities drove the development of S&T capability goals and objectives. Those goals and objectives constitute the heart of this *Roadmap* and comprise a guide for future Federal, academic, and industrial S&T efforts and for international collaborations. Sidebars highlighting some successful S&T programs have been included in this report as well, to demonstrate where current programs are working to address some of the gaps identified and where expansion of ongoing activities would be beneficial. While the intent was to create a *Roadmap* that applies to a variety of biological agents, it should be noted that many components and highlighted programs in this *Roadmap* apply in particular to agents that pose the difficult challenge of persisting in the environment (e.g., *Bacillus anthracis*).

Purpose and Scope

The strategic goals and objectives presented in this *Roadmap* aim to focus Federal S&T efforts with the goal of enhancing operational decisions at various phases of the response and recovery process. This *Roadmap* complements, but does not duplicate, ongoing S&T collaboration in the area of Biosurveillance and Medical Countermeasures (MCM) that address S&T needs for protection against the occurrence of a biological incident; initial detection, diagnosis, or confirmation of a biological agent; prediction of the occurrence of or forecasting the impact of a biological incident or the development and use of MCM.

Response and Recovery				
Crisis Management		Consequence Management		
^	^	Remediation/Cleanup			Restoration/Re-occupancy
Notification	First Response	Characterization	Decontamination	Clearance	^
Initiate first response activities, including notification of proper authorities	Operational Coordination Law enforcement, intelligence, and investigative response	Develop/ implement strategies for characterization in facilities and the outdoors	Decontaminate outdoor areas and/or buildings	Provide guidance for determination of effectiveness of decontamination	Provide guidance for re-occupancy and reuse criteria and goals
	When and how to distribute medical countermeasures	Implement strategies and procedures to identify, stabilize, and maintain infrastructure and property	Decontaminate wide areas		Provide guidance for controls to implement, reduce, mitigate any potential exposures or future incidents after re-occupancy
	Recommend staying-in-place or evacuation		Implement required capabilities for sustained environmental decontamination operations		
Develop a public-engagement campaign	Recommend quarantine/isolation/social distancing	Determine requirements and methods to protect natural and cultural resources			Implement public messaging to instill confidence in the public and workforce that re-occupancy is safe
Evaluate Threat Credibility	Implement transportation restrictions		Implement decontamination waste handling requirements		
	Provide safety and health guidance and protections to impacted first responders and citizens	Implement strategies and means to contain and mitigate the spread of contamination and eliminate sources of further distribution (e.g., insecticides for flies)	Decontaminate critical infrastructure		Implement measures to retain, maintain and improve the economic vitality of a region
	Issue guidance on personal hygiene or decontamination				
	Provide support for mass casualty				Implement long term health treatment, intervention and surveillance strategy
	Establish mass medical treatment facilities				
	Implement modified standards of care				

Figure 1. Key Response and Recovery Decisions

Chapter 2
Capability Goals to Support Critical Response and Recovery Decisions

This *Roadmap* presents the S&T needs to support the following critical operational decisions that may be required at various phases of response and recovery following confirmation of a biological incident:

- Develop/implement strategies for characterization both indoors and outdoors
- Determine when and how to distribute medical countermeasures[4]
- Recommend staying-in-place or evacuation
- Recommend quarantine, isolation, or social distancing
- Implement transportation restrictions
- Provide safety and health guidance and protections to impacted first responders and the public
- Implement strategies and methods to contain and mitigate the spread of contamination
- Provide guidance for determination of effectiveness of decontamination
- Provide guidance for re-occupancy and reuse criteria and goals
- Decontaminate outdoor areas and/or buildings
- Implement decontamination and waste handling requirements
- Provide guidance for controls to implement, reduce, mitigate any potential exposures or future incidents after re-occupancy

The high-level goals below describe relative end-states that will enable decision making, while the objectives and accompanying sub-objectives identify scientific knowledge and technological needs to achieve those end states.

Goal 1: Characterize the extent of the incident to reduce exposure and save lives.

Biological incidents must be adequately characterized in order to enable effective decision making. Validated methods for environmental analysis can inform decision makers of the extent of an incident and guide actions such as recommending evacuation or staying-in-place or determining the extent to which MCMs are needed. This capacity requires high-throughput methods to support surge capacity for wide-area and large-scale incident characterization; coordinated sample-collection and analysis capability to support multiple agencies during response and recovery operations; and real-time modeling tools to support biosurveillance operations, including a capacity to incorporate post-incident characterization data to identify potentially contaminated areas and guide response and recovery operations.

---

[4] The following link provides the most current FDA information on drug therapy and vaccines to prepare our country for possible bioterrorism attacks:
http://www.fda.gov/Drugs/EmergencyPreparedness/BioterrorismandDrugPreparedness/default.htm

Integrated Consortium of Laboratory Networks (ICLN)

The ICLN is a multiagency effort chaired by the Department of Homeland Security (DHS) to bring together information, operations, and strategies from different laboratory systems for timely response to major incidents; participating networks include the Centers for Disease Control and Prevention's (CDC) Laboratory Response Network (LRN), the National Animal Health Laboratory Network (NAHLN), the National Plant Diagnostic Network, the Food Emergency Response Network (FERN), the Environmental Response Laboratory Network, the Veterinary Laboratory Investigation and Response Network (Vet-LIRN) and the Department of Defense (DOD) Laboratory Network. The ICLN is a forum to facilitate communication and collaboration to build relationships and tools to support a more effective integrated laboratory response during emergencies (e.g., terrorist attacks, natural disasters, and disease outbreaks). ICLN activities include work to harmonize rapid-screening protocols for a pathogen outbreak, develop assay-sharing and testing reciprocity policies, enable transfer of laboratory data between networks, promote network compatibility/interoperability, and develop information-sharing policies for surge-capacity demands.

Rapid Viability Polymerase Chain Reaction (RV-PCR)

The RV-PCR method was developed for detection of live *B. anthracis* spores in surface wipe, air filter, and water samples. It has been optimized to achieve a limit of detection of 10 to 99 *B. anthracis* spores per sample. The method is most useful during the recovery phase of the response because it allows for rapid and high-throughput sample analysis to determine presence or absence of viable (live) *B. anthracis* spores (in the presence of a large number of inactivated/dead spores). The method supports rapid response and recovery following a *B. anthracis* contamination incident. Products from the research efforts can be accessed at www.epa.gov/sam/BAPROTOCOL.pdf and www.epa.gov/sam/.

Protocol for Detection of *B. anthracis* in Environmental Samples During the Remediation Phase of an Anthrax Event

This step-by-step protocol (http://cfpub.epa.gov/si/si_public_record_report.cfm?dirEntryId=247752) was developed for EPA's Environmental Response Laboratory Network and Water Laboratory Alliance to support remediation decisions. The broad promulgation of this protocol has effectively increased laboratory capacity to analyze anthrax samples during a wide-area event. The protocol includes adaptation of many procedures of the CDC's LRN protocols which will lead to increased confidence in sample analyses.

Objectives to achieve characterization of the extent of the incident and to reduce exposure include:

1. Establish the location(s) of the confirmed biological agent in the environment.

   a. Identify and verify performance for environmental sample collection, preservation, transport, preparation technologies, and protocols for high priority agents that can be rapidly adapted for a large-scale incident;
   b. Establish appropriate test and evaluation capability for rapid environmental contamination detection technologies to ensure reliable and consistent performance across the response and recovery operations; and
   c. Develop integrated sampling strategies, guidance, and training to support capabilities for environmental sampling and analysis.

2. Integrate incident characterization data into biosurveillance situational awareness modeling tools.

   a. Develop protocols and technologies to rapidly assess agent critical characteristics, spread potential (including secondary aerosolization), and environmental persistence;
   b. Develop capability to integrate rapid field-to-lab and lab-to-field data and results interpretation into biosurveillance modeling tools; and

c. Ensure that biosurveillance situational awareness modeling tools provide capability to rapidly forecast incident physical perimeters, estimate risk of environmental exposure, translate and integrate analytical, intelligence, and clinical surveillance data.

Goal 2: Effectively communicate to reduce the impacts of a biological incident.

Effective response and recovery operations require Federal, state, and local leaders to have the tools and messages they need to rapidly and effectively communicate with each other, the public, the media, and the response community during a biological incident. Effective communication is critical to quickly and accurately convey information that can mitigate the consequences of an incident; provide consistent guidance to the public on what immediate actions they should take to protect themselves; and provide guidance to the public as to when it is safe to return to affected areas. Because information travels quickly through social media, leaders at all levels of government need to be aware of the challenges and opportunities that the modern media landscape may present during an incident. For example, quick communication of an incident through social media might require rapid, high-consequence decisions (such as ordering evacuations, quarantines, release of MCM stockpiles) early on and in the face of high uncertainty. However, effective risk communication plans using social media and traditional media (e.g., radio and TV) can be prepared before an incident, pulled "off the shelf," adapted to fit the specifics of an incident, and used to quickly disseminate life-saving information to the public. This information can be tailored and delivered to targeted audiences based on their locations and circumstances. Additionally, decision makers can use social media to obtain feedback from people in the affected areas, which may be useful to assess and improve response and recovery operations.

Risk communication during an incident requires previously verified communication systems, interoperability, and focus-group or other testing of message content. Questions arise around how best to distribute information and what types of information to distribute, and how that information will be received. Building a foundation in the social science of effective communication is critical to creating effective media communication mechanisms and leveraging existing social media networks. Regardless of the type of information, science-based knowledge products and improved technologies to assess consequences and impacts of interventions will support the implementation of informed, coordinated, public messaging campaigns.

Generation of Risk Communication Best Practices

The EPA, in conjunction with county governments and academic institutions, is conducting multi-year studies of risk-communication best practices and is eliciting stakeholder and public input to determine risk communication needs and test pre-scripted messages. To date, the best practices for risk communication themes that emerged from this research are being used to create pre-scripted messages based on a water contamination event to be used during listening sessions with public consumers. Project outcomes will help develop risk-communication guidelines for post-incident decontamination and clearance activities following an intentional biological environmental contamination.

The EPA, in conjunction with the Pueblo City/County Health Department of Colorado, is conducting a three-year study of effective risk-communication practices during the remediation phase of a biological contamination event. The objective of this research effort is to determine the preferred mechanism of message delivery that makes it more likely that risk communications will be trusted and understood by target audiences, as well as examine risk tolerance, risk perception, and the use of social media.

Anticipating the Public's Questions during a Water Emergency

The EPA has conducted research to identify information the public will most need during a major intentional water-contamination incident. The study involved collecting information from both utility professionals and members of the public in four large metropolitan areas across the United States. The combined list of 400 questions identified during the study can be sorted into five overall question categories including (1) details about the incident and who is affected, (2) issues regarding exposure to the contaminant, (3) actions people can take to protect themselves and others, (4) acceptable uses of water and availability of alternative water supplies, and (5) response and recovery. Findings are presented in a report entitled *Need to Know: Anticipating the Public's Questions during a Water Emergency* (EPA/600/R-12/020).

Emergency Management Modernization Program (EM2P)

The Joint Project Manager-Guardian is one of the 6 Joint Project Management Offices under the Joint Program Executive Office for Chemical and Biological Defense (JPEO-CBD) is undertaking an effort to design, procure, field, train and sustain an emergency management capability for Army installations and assigned personnel. The system will provide an integrated, all-hazards Emergency Management system and enable a common operating picture, mass warning and notification, and communication with Enhanced 911.

Communicating Effectively with the Public during an Anthrax Emergency

Surveys, focus groups, and expert panels have guided the development of materials that could be used to communicate with the public during an anthrax emergency. These activities are the result of partnerships involving a host of partners, including CDC, Food and Drug Administration (FDA), EPA, Association of State and Territorial Health Officials (ASTHO), National Association of County and City Health Officials (NACCHO), National Environmental Health Association (NEHA), Association of State Drinking Water Administrators (ASDWA), and American Academy of Pediatrics (AAP), as well as state and local health agencies. CDC has used this feedback to create fact sheets, videos, social media, and web pages that aim to answer questions that members of the public would likely ask during an emergency and advise them on what they need to do to protect themselves and loved ones.

Objectives to improve communication throughout all phases of response and recovery include:

1. Enhance response communications through leveraging existing or developing new technology.

   a. Characterize the performance of existing technologies and develop innovative uses and applications that address recognized challenges in disaster communications;
   b. Develop geo-targeting technologies to support customized, response critical information dissemination based on location of recipient;
   c. Develop geo-targeting forms of communication to improve efficacy and timeliness of delivering messages to responders, the public, the media, and decision makers; and
   d. Ensure communication system interoperability.

2. Use risk communication research to develop appropriate messages and means of dissemination to all stakeholders (domestic and foreign), including decision makers, first responders, the public, and the media.

    a. Conduct risk communication and risk perception research to develop public engagement campaigns that effectively address the uncertainties of public health threats;
    b. Develop a fundamental understanding of what sources and forms of information are most trusted and well-received by the public, and what their expectations for information are;
    c. Develop messages that incorporate findings from risk communication research, including through the use of comprehensive websites, booklets, or videos, to engage the public in educational activities related specifically to biological response and recovery pre-incident;
    d. Develop message dissemination means that take into account diverse populations, including those with special needs, during an incident;
    e. Develop a medical provider messaging campaign, and networks for sharing information, for patient decontamination, triage, treatment, and management; and
    f. Develop methods to counteract the dissemination of erroneous information by unofficial sources.

3. Develop methods and algorithms to determine public understanding and response actions based on public messaging.

    a. Conduct basic R&D on public understanding of messaging and develop methods to assess a person's understanding of message content;
    b. Gather data to forecast and model how information is disseminated after a person receives a mobile alert;
    c. Design tabletop exercises to evaluate the effectiveness of messages and strategies to disseminate messages;
    g. Develop methods to gauge information dissemination efficacy to affected populations, including improved speed, accuracy, and overall response, due to message content and mechanism of communication; and
    h. Develop metrics to evaluate public messaging effectiveness.

4. Develop the social science approaches to understand and interpret response outcomes and amend messaging campaigns.

    a. Conduct studies to evaluate public confidence in messages and predict anticipated response to messages based on various trust levels;
    b. Identify and understand mechanisms to reduce the potential for failure in public health and safety response as a result of messaging; and
    c. Conduct studies to understand social outcomes of risk communication and messaging campaigns, including social vulnerability, response, and resilience to catastrophic biological incidents.

Goal 3: Accurately assess risk of exposure and risk of infection.

Risk assessment can be used to determine the association between the hazardous characteristics of a known infectious or potentially infectious agent or material with the activities that can result in an individual's exposure to that agent or material. Risk assessment also evaluates the likelihood that such exposure will cause an infection, the outcomes associated with an infection, and the likely impact of MCM. There is a need for more reliable information on the fate and transport of microorganisms in the multitude of environmental matrices, including food, and conditions encountered in both indoor and outdoor settings. A solid understanding of the factors that influence the persistence and abundance of microorganisms and their dissemination, and the impact of potential mitigation methods (e.g., insecticidal treatments to prevent the redistribution of vector-borne pathogens such as Japanese encephalitis virus spread by mosquitoes or enteric bacteria spread by flies) is needed.

> **Scientific Program on Reaerosolization and Exposure (SPORE)**
>
> The Scientific Program on Reaerosolization and Exposure (SPORE) is an interagency collaboration between DHS S&T, DOD Defense Threat Reduction Agency (DTRA), Department of Health and Human Services (HHS) Assistant Secretary for Preparedness and Response (ASPR), CDC Anthrax Management Team, and the EPA National Homeland Security Research Center (NHSRC, which serves as the interagency lead). The purpose of the program is to understand reaerosolization to inform response decisions. Initial projects are providing empirical data on the forces required to initiate reaerosolization from selected urban surfaces, evaluation of parameters that influence reaerosolization (surface roughness, humidity, spore preparation, etc.), and comparison of surrogates (both biological and inert) to *B. anthracis*. Interagency SPORE partners are planning and reviewing future proposals among partner agencies to ensure a focused and unified research approach.

> **Leveraging the North American Soil Geochemical Landscapes Project (NASGLP)**
>
> Through the U.S. Geological Survey (USGS) NASGLP, over 4,800 soil samples were collected across 48 states and uniformly analyzed for more than 40 major and trace elements. EPA and USGS teamed up to expand the project to examine the presence of naturally occurring high-priority pathogens. Presence/absence data will be mapped using geographic information systems (GIS) linking location to geochemical properties of the soil, ambient meteorological conditions, soil moisture content, and land use. A graphical representation of areas within the United States that may have high probability of naturally occurring biothreat agents is critical to decision-makers in determining if a detected constituent is part of the naturally occurring environment or a contaminant associated with an accidental or intentional release.

The following objectives provide the foundation from which to build a robust, quantitative risk assessment capability to enhance response and recovery:

1. Develop approaches and algorithms to assess risk of infection from environmental exposure, including food and water, to biological agents.

   a. Generate accepted methods to establish estimates of agent infectious dose as well as critical toxic levels that correlate to a clinical outcome;
   b. Develop knowledge of exposure pathways that account for incident scenario and unique properties of the agent;
   c. Generate plant and animal models to understand and forecast clinical outcome as a function of exposure routes and concentrations; and
   d. Integrate exposure research results with existing data on industry and consumer practices into MCM selection and distribution efforts;

2. Develop reliable estimates of risk of environmental exposure for a multitude of environments, matrices, and conditions associated with wide area release scenarios.

   a. Establish key scenarios to guide fate and transport research investment, including the impact of fluids (air and water) on the spread of contaminants;
   b. Conduct research to assess microbial organism's fate, transport, and temporal natural occurrence (background soil, water, food, and aerosol levels) and geographic distribution of biological agents;
   c. Develop tools to monitor changes in agent fate and transport over spatial and temporal variation;
   d. Develop degradation algorithms to predict persistence and examine fate during mitigation and recovery procedures (e.g., food, water and wastewater distribution and collection systems);
   e. Determine the risk of exposure to residual contamination as a function of environmental conditions and types of contaminated materials/surfaces to inform recovery decisions;
   f. Evaluate the behavior of biological contaminants in fatality management operations (e.g., burial, cremation) and potential environmental release as a result of those fatality management operations; and
   g. Utilize exposure research results to develop consensus methodology to determine when areas should be repopulated.

3. Develop reliable estimates of risk to humans, animals, and plants through various exposure and transmission routes.

   a. Establish key scenarios to guide research investment on the routes of transmission for emerging and ill characterized zoonotic, non-zoonotic, and foreign plant and animal disease pathogens;
   b. Conduct research to assess transmission mechanisms; and
   c. Apply established models for contaminant distribution in water, in potential edible products, and on insects, to make quantitative estimates of risk and characterize the certainty of those risk estimates.

Goal 4: Reduce the risk of exposure and/or infection.

Methods to reduce the risk of exposure or infection are essential for rapid recovery from a catastrophic biological incident. Risk-reduction strategies include actions such as quarantine, transportation control, social distancing, evacuation, distribution of MCMs, mitigation of the spread of contamination (e.g., removing contaminated water, insecticidal treatment of vector insects, or elimination of an infected flock or herd), and decontamination. An assessment that provides quantitative estimates of risk and characterizes the certainty of those risk estimates will inform decision as to which risk-reduction strategies will be most effective. For example, current guidance on the use and application of existing decontamination methods and strategies is limited in application and may not be appropriate for all potential biological agents. Development and integration of new decontamination methods and technologies requires standardized approaches to demonstrating safety and efficacy.

Decontamination Family of Systems (DFoS)

The Joint Program Executive Office for Chemical and Biological Defense (JPEO-CBD), Joint Project Manager Protection, at the Department of Defense has established programs under the DFoS umbrella that will improve decontamination processes by developing, maturing, and fielding materiel solutions to mitigate the hazards associated with chemical, biological, and non-traditional warfare agents and radiological contamination on personnel, equipment, fixed facilities, terrain, vehicles, ships, and aircraft. Programs currently within the DFoS include: Joint Sensitive Equipment Wipe, General Purpose Decontaminants, and Contamination Indicator Decontamination Assurance System.

Advances in Decontamination for Biological Agents

The Bio-response Operational Testing and Evaluation (BOTE) Project was a multi-agency effort designed to operationally test and evaluate response to a biological incident (release of *B. anthracis* spores) from initial public health and law enforcement response through environmental remediation. The effort was led by the EPA, DHS S&T, and CDC and included participation by the DOD, the Federal Bureau of Investigations (FBI), and the Department of Energy National Laboratories. The two-phase effort was designed to operationally assess the effectiveness, efficiency, and cost implications of decontamination approaches that advanced the cleanups following the 2001 anthrax incidents or that have been developed since that time. The coordinated response, including implementation of site remediation activities, was conducted and assessed during the second phase.

The EPA's Homeland Security Research Program and the Chemical, Biological, Radiological, and Nuclear (CBRN) Consequence Management Advisory Team continue to partner to advance remediation capabilities to promote community resilience after a biological incident of significance. Products from the research efforts can be accessed at www.epa.gov/nhsrc.

Under the DHS S&T Wide Area Recovery and Resiliency Program (WARRP), the EPA and Sandia National Laboratories have partnered to develop a decontamination decision-support tool (DeconST) to aid in the assessment of remediation strategy options. DeconST incorporates a systems-thinking perspective on decontamination, considering the relationship between sampling, decontamination, and waste-management capabilities. The tool enables the use of the latest decontamination-related research and capability advances to be used to support decision making, aiding the transfer of information to decision makers. From WARRP, DeconST is being transitioned to the EPA for maintenance and further development; the tool is also to be incorporated as part of the digital dashboard tool set (TaCBoaRD) that is being developed under the DTRA's Transatlantic Collaborative Biological Resiliency Demonstration (TaCBRD).

Objectives to achieve risk reduction include:

1. Ensure effective risk reduction strategies, including decontamination, waste management, contaminant control, and reaerosolization control, for a variety of biological threats and scenarios.

   a. Generate response plan guidance, including decision support tools and methodologies, that reduces risk by accounting for scenario and agent characteristics;
   b. Conduct research to assess the impact of existing mitigation technologies and protocols;
   c. Develop technologies and guidance for mass human, animal (including household pets), and plant decontamination or destruction;
   d. Conduct research on social and economic variables that promote or impede compliance capacity based on risk perception;
   e. Identify consistent and efficient risk reduction methods that are readily available, inexpensive, compatible with sensitive surfaces and materials, and environmentally friendly;
   f. Develop new approaches to decontamination and evaluate efficacy for new and existing decontamination approaches and procedures;

g. Develop methods to determine efficacy of risk reduction activities on the risk of exposure as a result of wide area incidents that result in a loss of property and infrastructure due to denial of use;
h. Develop outdoor (including surface waters), indoor, and water-distribution and sewage-waste-water-system risk-reduction strategies to reduce impacts and protect environmental and public health following a biological incident;
i. Develop science to support ventilation designs, including transport vehicle protective ventilation designs that provide for expedient and cost-effective isolation and negative pressure isolation rooms for use during emergencies;
j. Enhance individual survival by reducing the physical burden of personal protective equipment, integrate this capability into the responder's ensemble and use dynamic multifunctional materials that respond to threats and continually optimize between protection and burden;
k. Develop cost-effective designs to mitigate the potential occupational hazards involving airborne, vector-borne, direct contact transmission, and droplet-disseminated infectious diseases, including technologies to provide exposure protection for emergency medical professionals from airborne and surface contamination during response; and
l. Develop cost-effective and easily applied fixatives that effectively mitigate contaminant spread depending on incident scenario.

2. Develop a strong scientific basis for recommending population infection prevention measures (quarantine, isolation, and social distancing)

a. Perform studies to ascertain the risk associated with population movements during and after multiple disasters;
b. Develop alternative refuge options that enable access to clean air, food, and water while reducing the risk of continued exposure;
c. Develop expedient isolation methods and improved isolation capabilities to reduce exposure; and
d. Conduct systems-based analysis and develop evidence-based, optimized MCM distribution constructs, such as User-Managed Inventory (UMI), to supplement or enhance existing distribution mechanisms (e.g. stockpiles and points-of-distribution).

3. Implement risk reduction strategies to mitigate exposure from known routes of transmission, including reaerosolization

a. Conduct research and develop technologies to mitigate transmission of emerging and ill characterized zoonotic and foreign plant and animal disease pathogens;
b. Develop technologies to mitigate secondary reaerosolization of agents (i.e., containment of contamination) and reduce contamination spread following an initial release; and
c. Develop protocols and implementation guidelines to reduce the transport of agents by contact with surfaces, including fomites, and by transmission from insect vectors.

Goal 5: Manage biological waste following a catastrophic incident.

A key component of the response and recovery process is the proper management of wastes generated as a result of the initial contamination incident, initial response activities, natural processes that occur in the aftermath of the incident (e.g., precipitation), and the mitigation and decontamination operations that occur in the medium- and long-term phases of the response and recovery. Waste management includes methods and protocols for reduction, treatment, and disposal of agent-contaminated waste, and must be done in coordination with the law enforcement, intelligence collection, and investigative responses. Many of the knowledge gaps in the waste management area involve uncertainties associated with application of existing well-established technologies (e.g., waste segregation, recycling, landfilling, incineration, composting) to wastes that may have trace levels of unconventional contaminants (e.g., *B. anthracis* bound in building materials). At times these unconventional wastes may be generated in exceedingly large quantities (e.g., animal carcasses from a foreign animal disease outbreak or plants from contaminated nurseries or forests). There is a need to fill knowledge gaps relating to the remediation and cleanup of water distribution systems and wastewater that may be generated as a result of cleanup activities, run-off, or other incident-specific actions. Furthermore, more needs to be known about the agent-specific behavior and environmental release of biological contaminants –including, for example, spore-forming bacteria or soil-tolerant viruses—in fatality management operations, including burial and cremation. There is also a need to develop innovative approaches to minimize contaminated waste (by, for example, vaccinating animals in an outbreak so they don't need to be euthanized) and to enhance erosion control and water treatment methods to reduce the spread of pathogens by surface-water transport.

The Human Remains Decontamination System (HRDS) family of systems

The HRDS family of systems will provide the capability to protect personnel handling and processing human remains associated with a CBRN event. The HRDS will contain CBRN-contaminated remains from the point of fatality to the Mortuary Affairs Contaminated Remains Mitigation Site, reduce the hazard/eliminate the hazard from contaminated remains, and contain remains during storage and transportation via military airlift and/or commercial aircraft. HRDS consists of three major systems: the Contaminated Human Remains Pouch, the Contaminated Human Remains Transfer Case, and the Remains Decontamination System.

I-WASTE

The EPA, with financial support from the DHS and U.S. Department of Agriculture (USDA), developed the I-WASTE Tool over the past several years. The I-WASTE Tool contains estimators for various types and volumes of waste. The tool also provides location and contact information for treatment/disposal facilities, and health and safety information to ensure public and worker safety during the removal, transport, treatment, and disposal of contaminated waste and debris.

The I-WASTE Tool is available at www2.ergweb.com/bdrtool/login.asp.

Objectives to meet waste management goals include:

1. Manage and minimize the amount of waste generated from responding to all-hazards incidents.

   a. Develop tools to manage wastes as part of an integrated response and recovery operation, including prioritization of infrastructure cleanup, waste segregation, recycling, staging, storage, treatment, transportation, disposal of wastes, and assessing the impact of decontamination decisions on waste management practices;
   b. Adapt existing waste management (e.g., treatment) technologies in a mobile deployment setting to minimize contaminant spread and to reduce transportation and disposal costs (overall waste management costs);

    c. Develop methods to determine the level of contamination present in medical waste and hospital consumables to guide disposal and/or recycle, reuse of materials; and

    d. Develop guidance for agent and scenario specific concerns regarding the handling of decontamination wastewater and wastewater arising from normally occurring activities.

2. Develop methods and criteria for selecting waste management approaches for agent-contaminated waste.

    a. Evaluate human and environmental health risks associated with use of different waste management methods to inform incident-specific decisions;

    b. Develop scientific knowledge to assess cost/benefit analysis of waste management technologies, including long-term human and environmental health impacts;

    c. Research social acceptance and develop conflict resolution processes for gaining acceptance of waste management decisions from the public;

    d. Develop knowledge and performance evaluation measures of existing waste management (e.g., treatment and disposal) technologies for known and emerging biological agents for ease of technology selection for use in an incident;

    e. Generate mitigation strategies for wastewater collection and treatment, as well as prevention of drinking water system contamination during waste management actions, to support environmental decision making for sustained operations;

    f. Develop strategies and methods to mitigate the potential release of biological agents during all waste handling operations (e.g., effective containment during operations such as removal of materials from sites, temporary storage, transport, treatment, and disposal);

    g. Develop design criteria for constructing new treatment/disposal facilities; and

    h. Develop sampling strategies for minimizing the number of environmental samples for characterizing wastes prior to transportation.

3. Develop methods and criteria for evaluating and determining appropriate disposal of contaminated human and animal remains, and infested plants.

    a. Develop scientific knowledge to compare human and environmental health risks associated with treatment and disposal methods for human and animal remains and infested plants;

    b. Develop science to support long-term environmental and human health impacts of disposal methods; and

    c. Develop environmental criteria (e.g., soil type, depth to ground water, and distance to surface water) for onsite burial of animal carcasses.

Conclusions

A catastrophic biological incident would force leadership at all levels of government to make rapid, high-consequence decisions in the face of numerous uncertainties. This *Roadmap* identifies knowledge and technology gaps that currently impede the ability of relevant officials to act quickly and decisively on the basis of sound science and reliable evidence, and presents goals and objectives to speed the filling of those gaps. Although progress is already being made in these important domains, continued progress will require focused investments and coordination among Federal departments and agencies, academia, industry, and international partners as recommended in this roadmap.

Appendix I: Definitions

- *aerosolization, n*—the process of generating an aerosol; a process that generates small particles that can be carried in the air.
- *biological incident, n*—a natural or human-caused incident involving microbiological organisms (bacteria, fungi, and viruses) or biologically-derived toxins that pose a hazard to humans, animals, or plants (Draft Planning Guidance for Recovery Following Biological Incidents).
- *biosurveillance, n*—the process of gathering, integrating, interpreting, and communicating essential information related to all-hazards threats or disease activity affecting human, animal, or plant health to achieve early detection and warning, contribute to overall situational awareness of the health aspects of an incident, and to enable better decision making at all levels (The National Strategy for Biosurveillance).
- *biothreat agent, n*—any microorganism, virus, infectious substance, or biological product that may be engineered as a result of biotechnology, or any naturally occurring or bioengineered component of any such microorganism, virus, infectious substance, or biological product, capable of causing: (1) death, disease or other biological malfunction in a human, an animal, a plant, or another living organism; (2) deterioration of food, water, equipment, supplies, or material of any kind; (3) or, deleterious alteration of the environment (18 USC 175).
- *catastrophic biological incident, n*—a natural or manmade incident, including terrorism, involving microbiological organisms (bacteria, fungi, and viruses) or biologically-derived toxins that results in extraordinary levels of mass casualties or disruption severely affecting the population, infrastructure, environment, economy, national morale and/or government functions.
- *catastrophic incident, n*—any natural or manmade incident, including terrorism, that results in extraordinary levels of mass casualties, damage or disruption severely affecting the population, infrastructure, environment, economy, national morale and/or government functions (NIMS).
- *characterization, n*—the process of obtaining specific information about a biological agent, such as its identity, genetic composition, formulation, physical properties, toxicological properties, ability to aerosolize, and persistence, and about the nature and extent of contamination of the agent, such as locations or items contaminated and the amount of contamination (Draft Planning Guidance for Recovery Following Biological Incidents).
- *cleanup, n*—the process of characterizing, decontaminating, and clearing a contaminated site or items, including disposal of wastes. Cleanup is a synonym for Remediation. Generally occurs after Characterization and before Clearance (Draft Planning Guidance for Recovery Following Biological Incidents).
- *clearance, n*—the process of determining that a cleanup goal has been met for a specific contaminant in or on a specific site or item. Generally occurs after Decontamination and before Re-occupancy (Draft Planning Guidance for Recovery Following Biological Incidents).
- *community resilience, n*—the ability of a network of individuals and families, businesses, governmental and nongovernmental organizations and other civic organizations to adapt

to changing conditions and withstand and rapidly recover from disruption due to emergencies.
- *consequence management, n*—actions taken to maintain or restore essential services and manage and mitigate problems resulting from disasters and catastrophes, including natural, man-made, or terrorist incidents. Also called CM. Includes Remediation/Cleanup (i.e., Characterization, Decontamination, and Clearance) and Restoration/Re-occupancy activities.
- *contamination, n*—the process of making a material or surface unclean or unsuited for its intended purpose usually by the addition or attachment of undesirable foreign substances. Used in this document to describe building, water, and outdoor exposure and external human, animal, and plant contamination.
- *crisis management, n*—measures to identify, acquire, and plan the use of resources needed to anticipate, prevent, and/or resolve a threat or an act of terrorism. It is predominantly a law enforcement response, normally executed under Federal law. Also called CrM. Includes Notification and First Response activities.
- *critical infrastructure,* n—systems, assets and networks, whether physical or virtual, so vital to the United States that the incapacity or destruction of such systems and assets would have a debilitating impact on security, national economic security, national public health or safety or any combination of those matters.
- *decontamination, n*—decontamination is a process that makes an item, instrument or device safe to handle and can be accomplished by cleaning with detergent and water, or cleaning in combination with disinfection or sterilization.
- *emergency responder, n*—includes Federal, state, local, and tribal emergency public safety, law enforcement, emergency response, emergency medical, including hospital emergency facilities, and related personnel, agencies, and authorities. See Section 2 (6), Homeland Security Act of 2002, Pub. L. 107-296, 116 Stat. 2135 (2002). Also known as *Emergency Response Provider* (NIMS).
- *emergency response, n*—the performance of actions to mitigate the consequences of an emergency for human health and safety, quality of life, the environment and property. It may also provide a basis for the resumption of normal social and economic activity.
- *exposure risk, n*—the probability of being exposed to an infectious agent, chemical intoxicant, or radioactive substance resulting in a degradation of health.
- *evacuation, n*—organized, phased, and supervised withdrawal, dispersal, or removal of civilians from dangerous or potentially dangerous areas, and their reception and care in safe areas (NIMS).
- *evidence-based, adj*—medicine, pertaining to the conscientious, explicit, and judicious use of current best evidence in making decisions about the care of individual patients.
- *first response, n*—actions taken immediately following notification of a biological incident or release. In addition to search and rescue, scene control, and law enforcement activities, first response includes initial site containment, environmental sampling and analysis, and public health activities, such as treatment of potentially exposed persons (Draft Planning Guidance for Recovery Following Biological Incidents).
- *hazard, n*—something that is potentially dangerous or harmful, often the root cause of an unwanted outcome; a danger or peril (NIMS).
- *high consequence decision, n*—a decision that could result in major disruption or has important health or economic risks. Examples of high consequence decisions in the

context of this *Roadmap* include: ordering evacuations, quarantines, and release of MCM stockpiles.
- *infection risk, n*—the risk to an individual of developing an infection following exposure to a pathogenic organism (Virus, Bacteria, Fungi, etc.).
- *intervention, v*—to involve oneself in a situation so as to alter or hinder an action or development.
- *jurisdiction, n*—a range or sphere of authority. Public agencies have jurisdiction at an incident within their area of responsibility. Jurisdictional authority at an incident can be political, geographic (for example, city, county, tribal, state, or Federal boundary lines) or functional (for example, law enforcement, and public health) (NIMS).
- *notification, n*—the process of communicating the occurrence or potential occurrence of a biological incident through and to designated authorities who initiate First Response actions. Generally occurs as the first step in a response to a suspected or actual biological incident (Draft Planning Guidance for Recovery Following Biological Incidents).
- *persistence, adj*—the ability of an agent to live or endure outside of the host and remain infectious.
- *reaerosolization, v*—when particles collected on surfaces are re-entrained into the air stream.
- *recovery, n*—the development, coordination, and execution of service- and site-restoration plans; the reconstitution of government operations and services; individual, private-sector, nongovernmental, and public assistance programs to provide housing and to promote restoration; long-term care and treatment of affected persons; additional measures for social, political, environmental, and economic restoration; evaluation of the incident to identify lessons learned; post incident reporting; and development of initiatives to mitigate the effects of future incidents (NIMS).
- *remediation, n*—the processes of characterizing, decontaminating, and clearing a contaminated site or items, including disposal of wastes. Cleanup (Draft Planning Guidance for Recovery Following Biological Incidents).
- *re-occupancy, n*—the process of renovating a facility, monitoring the workers performing the renovation, and deciding when to permit re-occupation. Generally occurs after a facility has been cleared but before occupants are allowed to return (Draft Planning Guidance for Recovery Following Biological Incidents).
- *response, n*—activities that address the short-term, direct effects of an incident. Response includes immediate actions to save lives, protect property, and meet basic human needs. Response also includes the execution of emergency operations plans and of mitigation activities designed to limit the loss of life, personal injury, property damage, and other unfavorable outcomes. As indicated by the situation, response activities include applying intelligence and other information to lessen the effects or consequences of an incident; increased security operations; continuing investigations into nature and source of the threat; ongoing public health and agricultural surveillance and testing processes; immunizations, isolation, or quarantine; and specific law enforcement operations aimed at preempting, interdicting, or disrupting illegal activity, and apprehending actual perpetrators and bringing them to justice (NIMS).
- *restoration, n*—the process of renovating or refurbishing a facility; bringing it to an acceptable condition using the optimization process to determine the appropriate use and associated clearance level at which occupants may return. Generally occurs after the

Clearance Phase but before occupants are allowed to return (Draft Planning Guidance for Recovery Following Biological Incidents).
- *risk, n*—the probability that a substance or situation will produce harm under specified conditions. Risk is a combination of two factors: (1) the probability that an adverse event will occur (such as a specific disease or type of injury), and (2) the consequences of the adverse event (Presidential and Congressional Commission on Risk Assessment and Risk Management, 1997).
- *risk assessment, n*—gathering and analyzing information on what potential harm a situation poses and the likelihood that people or the environment will be harmed. [The Presidential and Congressional Commission on Risk Assessment and Risk Management, 1997] A methodological approach to estimate the potential human or environmental risk of a substance that uses hazard identification, dose–response, exposure assessment, and risk characterization.
- *staying-in-place, v*—the act of remaining in a designated location in order to avoid further danger elsewhere and/or to limit negative impacts in the area of the emergency. Staying-in-place is often recommended in order to facilitate the distribution of MCM or aid to individuals. This is in contrast to "shelter-in-place" which means to take immediate shelter where you are—at home, work, school, or in between.
- *surge capacity, n*—a system's ability to rapidly mobilize to meet an increased demand, to rapidly expand beyond normal service levels to meet the increased demand in the event of large-scale disasters or public health emergencies.
- *transport, v*—the movement of contaminants through environmental media (e.g., air, soil, water, groundwater).
- *threat, n*—an indication of possible violence, harm, or danger and may include an indication of intent and capability (NIMS).

www.ingramcontent.com/pod-product-compliance
Lightning Source LLC
Chambersburg PA
CBHW081825170526
45167CB00008B/3543